MY CHILDHOOD PASSION

I want to make it really really clear that I am the last person you'd see driving a Prius.

I am a car guy, a race car driver, I LOVE burning fuel and rubber. When I was a kid my uncles and father raced stock cars at the local short track. I've wanted nothing more than to race since I was 6 years old.

I was about 15 when dad brought home a race car project for me. We worked and worked and wrecked and worked.

By the time I was 18 we were winning races!

I attended a vocational Highschool to learn the trade of Auto Technician. I attended college to learn the trade of Welding Technology.

I started a family young and racing had to take a back seat. I am a father, husband, and I wear a few other hats but, I never lost my passion for cars and speed.

No one enjoys pouring into a corner on the absolute edge of control, rolling through the apex with as much momentum as possible, an inch away from another maniac doing the same thing, as I do. I roll into the throttle as soon as possible, instantaneously adjusting input to manage forward bite and minimize tire spin, as my foot reaches the floor, I push hard asking for just a little more out of the maxed engine. The race built chevy small block screams in a rage of noise, heat, and horsepower. I prepare to dive into the next corner as deep as possible. I roll off the throttle smoothly, apply the brake with as much finesse as can be, just enough to settle the weight transfer, make the corner, and dance on the absolute edge of control again. UGH, I LOVE THAT SHIT! My whole body is tingling and I'm close to tearing up writing this.

Most recently, my father, son, friends, and I built a sportsman modified! I'm out of retirement and back on track. This is my ultimate form of meditation. When I am on track, I lose all thought, sense of self, and become another component, one with the car.

I can feel every sensation of its function in my being. It's as if the engine is my breath, my life force energy. The oil is my blood, pumping hot through our veins. Power is delivered through the crankshaft like its muscle in my legs. I feel the grip and slip of the tires in my gut.

Like people, tires start young and full of energy, build up and reach max potential, then honorably serve with all they have as they lose grip and life.

The RR is worn out, it has no more to give, I feel less and less grip every lap. It's dying but, it's giving all it has on the way out. I feel a clink, the LF upper ball joint nut is loose, I feel it as if it's my own wrist. After the race, my family, friends, and fans are my heart manifested to life as we connect and feel what we just experienced. As you can see, My whole life has been passionately burning fuel. Lots of fuel as inefficiently as can be.

WORK AND EXPOSURE
TO THOUSANDS
OF CARS.

I was also mechanic for 10+years and spent more time under cars, driving cars, critiquing cars, reviewing cars, and making cars fast than most. Every car I've ever repaired was finished with a test drive. Every test drive was an opportunity for me to feel the driving dynamics.

I was also fortunate enough to work for my friend Pat, who would let me drive his fun cars. Nissan 200z, new Escalades, Denali Pickups, BMW M6, V-12 Bentley, Grand Cherokee Trail Hawk, and a bunch of other fun whips. By far my favorite was his 2014 Porsche 911 50th Anniversary edition!

With the wide-body, dropped stance, and pure driving bliss it was my most joyful driving experience outside of a race track. I can't even get started, as the perfection of that car is an article in itself. He sold it for more than he paid brand new from Porsche. I decided right then and there that my next daily driver will be a 911, if and when my BMW E46 330i ever gives up. No One loves and appreciates the feel of a properly built sports car like I do.

MY DAILY

I drive an old Bimmer because I love RWD, manual trans, proper weight distribution, excellent brakes, and all the thoughtful details of German engineering.

She ain't bad to look at either. This car is so tough, has brought me so much joy, has been on such epic journeys, that our relationship also requires its own article.

After research on lap times, fuel economy, rear seat hip width, and many other attributes on a spread sheet. I decided on a prefacelift e46 330i with sport and winter package.

I was all set to road trip to virginia for the car I wanted and this car popped up local on craigslist. I bought it 7 years ago for $6000 and put over 125k HARD miles on it.

We rely on each other in work, play, and life. No other vehicle has survived 100k of my demanding driving. I have driven this car for 12 hours straight with no complaints. It's no old, tired, and beatup with 250,000miles on the clock but, it still brings a smile to my face.

But, this isn't about old Bimmers and expensive Porsches. I'm sharing this so you are absolutely sure that I am an authentic car enthusiast sharing my feelings about electric cars.

MY DREAM TRUCK

For the truck guys out there I also have an old school, big block, 3 speed, Suburban. We have 3 kids and a huge dog, so we need a minivan but, I figured neither one of us wants to drive a lumbering fat pig of a minivan or suv to work and back daily.

After much thought on possible options this is my answer. We both get a daily driver that makes us smile and I have an old Suburban for minivan duties when needed. Which also makes me smile.

I manifested the exact truck I wanted. The color, options, drive train, headlights, with seating for 9, and the towing package, this thing fell from heaven, and landed on eBay for $6,000.

 I wanted a huge truck that works as a minivan, a race car hauler, and can cruise to car shows. I'm also a sucker for the *Square Body Round Eye Chevy Trucks* that I grew up with. This thing is as far from "eco" as one can get. The tired old engine manages about 12mpg.

The build quality of an old Chevy truck is satisfying. The clunky mechanism of manual locks, the shifer linkage, highbeam swith on the floor. No modern vehicle has this much passenger and cargo room. The doors are heave to slam and my kids have to operate roll up windows. All of which makes me smile.

MY BOY

To add more fuel to the enthusiast fire I have to mention that my kid is a badass Kart racer. "Racin Mason" loves to burn fuel and rubber more than I do. He's more talented than me too!

Check him out https://www.facebook.com/RacinMason33

https://www.instagram.com/Racin_Mason33/

He has had the racing bug since before he could walk or talk.

An absolute natural on the track, winning his second time on the track. he went on to win all over the noth east in 2017 and won New York State championship in 2018.

Mason is different from most young JR racers in that he wrenches on his kart and makes his own adjustments.

The boy is all heart, desire, and effort. He goes hard every lap.

Mason took 2019 off to help me assemble the sportsman modified. I'm hoping to have a modified sorted out with good equipment to match his amazing talent when he's eligble to race a big car at age 14.

We haul the Kart all over the northeast with the Bimmer. I want to enjoy the road trip to the track.

I AM A HUGE FAN
OF THE RIGHT CAR
FOR THE JOB

Let it be clear that I am an unbiased car enthusiast. I love anything from old American V8s, rough and tough race cars, nimble European cars, and anything else that excites my senses. I've driven everything from an old 5 speed Saturn with an exhaust leak to a new Bently. If it's fun, boring, smart, dumb, well built, or junk. I call it what it is with no discrimination.

I really appreciate the right tool for the job. I wouldn't wear Nike ACG's to play basketball in and I wouldn't wear Jordans to hike in. In the same way I wouldn't cross the African safari in a full-size sedan and I wouldn't cruise the highway every day in a Toyota 4-Runner.

The 4 Runner is one of the most capable, dependable beasts one can buy. If I lived in Alaska or Australia and my life depended on traveling rough roads through wilderness then the 4 Runner is my whip. But, If I have a daily commute from the suburbs to the city and its mostly freeway I'll have a powerful, smooth comfortable sedan. Anything from a used V-6 Camry to an old BMW 5 series (value). I'd recommend the Camry for the average person who doesn't appreciate driving dynamics as much and just wants to get to work. I'd recommend the Bimmer to an enthusiast who appreciates the driving sensation, build quality, and is willing to pay more in maintenance or repair.

I drive cheap old vehicles. I have a car buying philosophy called "Fu*k a car note!" The whole dealership and car loan game is predatory. I have seen too many people commit to a car based on monthy payment and go upside down. It is quite common for people to owe more on their car than it's worth. This backward math is not a happy way to live. Especially if you wandered into a dealer with a rough idea of what you wanted and were sold something else. If you choose to spend thousands in depreciation, interest, lease, and fees that's fine. Just be aware of what you are doing with your finances. Make your own choice, don't let the industry make your choice for you.

For those of us who want the most transportation value for our money we have to use a different stragety. Usually, for the average person this means buying slightly used and driving it into the ground. Hopefully it is relaible enough to provide some note free years of happy transpotation.

This stragety requires research. We are most likely going to have this vehicle for a long time. It should make us happy. This is one of, if not, the biggest purchase after our home. Treat it as so.

Decide what you want from your vehicle. Economy, relaibilty, style, status, safety, fun, utility? What is your daily driving like? What vehicle would perfom best for what you need? What vehicle will make you smile every time you get in it?

For me, I feel happy about the timeless look of my car. The short front overhang and tight wheel to fender clearence. I feel happy shifting gears, working the clutch, and using the confident brakes.

For you it may be that it gets 40 mpg or it fits your family and dog. It may be simple relaibility and value that makes you smile. It may be comfort, the heated seats, the stereo system.

Whatever it is, figure it out, narrow down your search. Read reviews on edmunds, Car and Driver, Consumer Reports, Etc. Decide what model you want and research the different engines, trims, upgrades through model years. Decide exactly what you want as specifically as possible including make, model, years, trim, options, and colors.

Then hit the internet, find availibile examples, compare prices, go drive them. If you like it, narrow down the specific example you want. If not, research and keep looking for that daily source of happiness.

Cars have price ranges, try not to get caught up in small price differences that may cost you 5 years of happiness. If you want heated seats but the only model with heated seats is $500 more and 300 miles away, go get it! You'll be happy you did in the long run.

I'm not against traveling for the right car. It may take 2 days and cost $500 but, that's worth it if that's what it takes for a daily source of driving happiness.

If you want new and have money to burn in depreciation cool. If you want to minimize financial loss please do not be afraid of used.

You don't have to be a mechanic to say fu*k a car note. The first car I bought for my wife, then young girl friend, was a 10 yr old 100,000 mile Nissan Sentra for $2000.

I was driving a Caprice Classic at the time that we called "Big Dirty" So we named the Nissan "Clean Sentra" That car stayed in our family for another 10+ years and 125,000+ more miles. That whole time all we had to do was an alternator, clutch, tires, and brakes. We were happy to experince a decade of note free driving, 35mpg, room for car seats, and trusty reliabilty.

Do your research, do the math, find your happy spot. We spend every day with our car. Might as well make it enjoyable. Build a relationship. Give it a name. Cars like to be appreciated, thought of, and named. Trust me, the science behind the law of attraction is real. I have experimented with and reflected on this. Love your car and it will love you back.

MY LOVE GETS A NICE NEW CAR. CHOICES CHOICES CHOICES.

For wifey, we decided it was time for an upgrade. In June of 2018 we moved and her daily commute grew from 8 miles to 40 miles. Our trusty Nissan Pathfinder only gets about 14 MPG on a good day so it was a good excuse to upgrade. We wanted to stay in the high teens' price range. She was willing to go into the low 20s and wanted another SUV. We drove a few but, being a car guy they all felt overpriced, poorly designed, and inefficient. The 4 cylinder engines in todays SUVs are working way too hard.

Also, we don't really gain anything in today's soft, FWD car-based, SUVs over the sedan they are based on. In most cases the SUV is built on the same chassis as the equivalent sedan. The major difference is the SUV we get more cargo room, higher ride height, and more weight. All of which equals worse driving dynamics, and less efficiency, for a higher price.

WINTER DRIVING

Side note.... the inferior driving dynamics (handling, cornering, braking, maneuvering) of vehicles with a higher center of gravity like trucks and SUV's actually make them WORSE on slippery snow and ice. If a vehicle can stop shorter or turn faster on dry pavement, these abilities are only more prominent on ice and snow. If a vehicle has more advanced traction control and ABS, it is that much safer on the ice and snow.

We had a 2004 Chevy Suburban which would clumsily slide through an icy stop at 4mph while the brake pedal vibrated and the abs lights flashed. That same stop in my 2001 Bimmer was perfect as its ABS is far superior with multiple different functions for different situations. The dynamic stability control and ABS of an older but more advanced car saved me from what would have been definite accidents in a newer less advanced American vehicle. It hurts me to say it, but it's true, American companies are behind in build quality, technology, and electronic driving aids. They have been for decades.

The advantage of a truck or SUV in the winter is AWD or 4wd. This ONLY helps in accelerating quickly from a standing stop, beasting through deep snow, climbing a slippery incline, or yanking a stuck car free. Not that any of today's 4cyl cute utes could chain tow another car out of a ditch. A truck with good tires could. These situations are actually pretty rare in winter driving. I would know as I've spent 20+ years driving in Buffalo and across the state for school and work.

ROAD TESTING IN WINTER WEATHER

I've made sure to push every vehicle I've ever had (and some that weren't mine) to their limits in Buffalo's wintery conditions. I've been stuck in every one of them, sideways in every one of them, skidding in every one of them, and drove everyone into snow deeper than anyone in their right mind would. I've busted dough-nuts in unplowed parking lots, joyfully packing the wheels and suspension full of icy snow. . . . in every one of them. I've busted so many donuts with my kids that they don't even look up from their phones as I practice slideways figure 8's with my Bimmer in the local college parking lot.

When I had to drive 100 miles to college every week and 300 miles to work downstate for 3 years, I really learned how much better a car is generally than a truck in the snow. If I'm in a blizzard, on a highway and have somewhere to be, give me a nice handling car with snow tires; which gives me the stability and confidence to drive at 50-60mph kicking up fresh powder, in perfect control. I've had a lot of pickups and SUV's to haul my race cars and they were definitely not as stable on the highway, or quick to stop like a car.

The shorter wheelbase and snow tires on the old Pathfinder made it a beast in deep snow. I've had plenty of fun dominating winter with that truck. If there was already 6" on the ground, it's snowing hard, and I had to leave before the country roads were plowed, I'd obviously prefer to take the Pathfinder's ground clearance over my Bimmer with its sporty low ride height, but that situation really happens about 1-3 times a winter. I'm not going to suffer daily in safety, efficiency, and **fun** to drive a heavy, inefficient, SUV for 363 days in case of the other 2 days I have to challenge deep, unplowed, snow.

I also learned that quality snow tires are EVERYTHING. It makes such a huge difference on all vehicles that I'd say it is irresponsible to drive in a winter climate without them. A narrow, brand new, all-season tire may be okay to cruise around the suburbs but, if you are regularly driving over 55mph snow tires are a must.

All-season tires and car-based SUV's are similar in that they aren't exceptional at anything but, can do everything at below average. There are exceptions to this generalization such as the Porsche Macan and Michelin Pilot Sport A/S 3. But, most entry-level SUVs and all-season tires don't do anything well.

That being said, a smart AWD car like a Subaru, Lexus, Audi Quattro, AWD Bimmer with snow tires is an absolute **blast to drive** in the winter. AWD cars leave SUVs with the only advantage of higher ground clearance. This is only really necessary in the inner city side streets where they don't plow. Or if you don't want to shovel the end of your driveway and beast over the packed snow barrier from the street plow. If you want an SUV or truck, get one, just don't tell me you bought it because it's better in the snow. I drive 35 miles of country back roads to work and 90% of the cars I see off the road stuck in ditches are SUVs. If I had to bet I'd say that 100% of them are wearing all-season tires. Rant over, back to Mams's car shopping.

CHOICES CONTINUED

Jean loves the look of the "baby Range" and we drove one but, it felt to me like an overpriced Subaru with an indecisive transmission. I definitely didn't want the maintenance of a used Range Rover Evoque. Our kids were a bit tight in the back seat as it was about as roomy as a Subaru, not very.

As I was researching cars, trying to find something fun that would make her smile, I realized she is a perfect candidate for an electric car. Her round trip to work is 40 miles/day. Like most people, she doesn't care about feeling the road through excellent steering or pouring into an off-ramp at twice the suggested speed. She hates pumping fuel, and I hate to spend my time repairing our daily drivers. Most of all she'd save a bunch of money in fuel!

After driving a few other overpriced, underwhelming SUV's I was able to talk my love, Jean into considering an *electric car* with the main argument that she'd never have to pump fuel in February ever again and I'd never have to work on it.

The problem was that they are all so ugly. She's right, we drove the BMW i3 which was very well designed but she wouldn't be caught dead in something that looks like that. The Nissan Leaf is so hideous that even I wouldn't be caught dead in that. We can't afford a Tesla and Jean says they look like a cat. She LOVES our cat Gus but, she doesn't want her car to look like a cat.

Jean also wasn't willing to adjust to the one-pedal driving of an electric car. When you lift off the accelerator the car slows as if you are braking. Dont worry the brakes lights come on. Under normal coasting conditions regeneration kicks in to transfer the energy of moving mass into electricity. This electrity is produced by the energy of the cars momentum and recharges the battery.

Electric motors are so beautifully simple. When they aren't using energy to create power they can be spun by an outside power source to create energy. Some examples are the alternator under our hood or the monster turbines under Niagara Falls.

To me, one-pedal driving was something new to master. I also love the idea of a set of brakes lasting 100,000 miles but, Jean wants to coast.

THE ELECTRIC DECISION!

Enter the eGolf! It looks exactly like a regular VW Golf minus the tailpipe. It also has 4 *selectable* levels of regeneration. Used 2015-2016 eGolfs were going for $12-18k but, the range is only about 100 miles and they haven't sorted out the heat that well. I did some research and found that the new 2017 had a farther range, better heat, and electric car incentives put it in our price range. I also saw that they were selling like hotcakes in Scandinavia and the modern-day Vikings have life figured out. If the eGolf works in Finland, it will work in Buffalo, NY.

I never would have considred new but, the $10,000 in combined government incentives really helps the math! She decided on which trim and color she wanted and I searched. There weren't any available locally. We drove a regular 4cyl 2017 Golf at our nearest VW dealer. We took our 3 kids to make sure we all fit.

The Germans have interior ergonomics mastered. I really do think they measure the natural joint angles of the sitting human body to maximize room and comfort. We were also impressed with the sporty efficient driving dynamics paired with a refined quality feel. Most importantly Jean thinks its sooo cute!

THE PURCHASE

I found the trim and color we wanted in Newport, Rhode Island. I love to negotiate and confidently asked for the dealership manager over the phone. He wouldn't budge on the purchase price. This was a first for me but, he had the leverage of supply and demand. The few that were on the internet located within 500 miles of me were flying off the market. I let it ride for one day, c-alled him back, and tried to low ball him again. No luck, we gave in to full purchase price of $28,500. We chose to road trip for the car as opposed to getting it delivered. I wanted to see, drive, and sign in person.

We made it into a fun adult weekend getaway. Dinner, drinks, and people watching. We soaked in the sunset over the bay and finished with a fun night in a nice room.

The next morning we drove the eGolf. It felt just like the regular golf except it is *much quieter, smoother, and feels more stable.* We signed the dotted lines and took a 5yr 0% interest loan through VW and financed about $30,000 with taxes and fees.

We hauled it home on a u-haul trailer with Pat's GMC 2500hd Sierra Denali Diesel. That beast barely felt the load as we set the cruise at 77mph back to Buffalo.

I don't know why there weren't any eGolf's available locally at the time but if they were we would have gotten another $2500 off the purchase price from New York State on top of the $7500 check from the feds. *The government incentives are real!*

I sold the Pathfinder on craigslist for $5000 after a few tire kickers, we ended up selling it to a really fun lady that absolutely loved it. We dropped the $5000 + $7500 on our loan so we are already down to about 17,500 right off the bat.

Factor in the roughly $3000/year we are saving in fuel, oil, and maintenance of the old Pathfinder, the brand new *electric car should pay for itself in about 6 years!* New York state also offers a discount at tolls with a special green ezpass.

AFTER 2 YEARS OF OWNERSHIP

We have now owned our electric car for just under 2 years and put over 30,000 miles on it.

THIS IS THE BEST CAR WE HAVE EVER OWNED!

THIS IS THE SMARTEST PURCHASE WE'VE EVER MADE!

IT IS HEALTHY

No fuel or oil EVER! Jean hasn't stepped inside a gas station in 2 years. This may be a small factor in her health and *happiness*. Next time you are in a gas station, mindfully look at everything that's in there.

Besides spring water, it is a one-stop-shop for the poisoning of humanity. Especially in the hood where you can buy fuel, lotto, freshly fried chicken wings, a vape cartridge, dutch master flavored wrap, a pair of leggings, some hair weave, 3 loose Newport cigarettes, sugary drinks, and salty snacks in one smooth transaction.

I'm not judging, I smoked loosies for a long time, I lived that life but, facts are facts. Gas Stations are strategically set up in a way to trigger us into buying junk that doesn't support a healthy lifestyle. Moving on...

IT'S VERY LOW MAINTENACE

Whereas cars with an internal combustion engine (ICE) needs about 30,000 components, the same vehicle in electric (EV) needs just 11,000 parts, according to research from Goldman Sachs Group INC. That reduction in complexity has lowered the barriers to entry for the automotive market and caused a surge in new electric cars coming to the market. The simplified design and engineering mean a lot less to go wrong and a lot less for me to repair.

That is the main reason I would NEVER buy a hybrid. It defeats the advantages of fully electric cars. Hybrids have more weight and more stuff to repair by having an internal combustion engine (ICE) , everything needed to run it, AND an electric powertrain. Besides for the Prius and first gen Insight most early hybrids are half-ass afterthoughts with a battery installed where the spare tire would normally go. This was done as more of a sales gimmick than any notable improvement in the car. Today there are more hybrids built from the ground up on dedicated platforms. This includes plug-in hybrids with the ability to run on fully electric and or internal combustion but, *why bother when fully electric exists*?

- *Electric cars don't have fuel.* No fuel lines, no fuel pump, no fuel injectors, no fuel filter, or any EVAP equipment to cause a check engine light.
- *It has no engine!* No exhaust, no muffler, no O2 sensors to cause a check engine light.

- *No transmission,* no solenoids, no transmission fluid, no transmission lines to leak.
- It has a *ton of less moving parts*. It's beautiful how simple and efficient the design is.

We did have one problem after multiple charges in the rain. This caused some corrosion in the charging port. I would have been satisfied if they just cleaned and lubed the connection at the dealer but they insisted on replacing it. We had a loaner for a couple days while we waited for the part to arrive.

We took it to the dealer 3 more times for each 10,000-mile service. Its kind of funny because there are no belts to check, no oil to change, no filters to change, and the brakes see little use. The service reps just kind of shrug their shoulders. There isn't much to look at.

We *haven't spent a penny on maintenance.*

COST OF CHARGING

We *haven't spent a penny on gas.* We do pay our electric bill but, it is super cheap to run. I called my electric company and switched to a time of use plan. This is a free option and now we pay 6.5 cents per kWh from 8 am to midnight and 3.1 cents per kWh from midnight to 8 am. With electricity being half price at night we plug in the EV and set the timer in the infotainment center to postpone charge until after midnight. We can do this from our phone in bed for a monthly fee but we declined.

We can manage about 5 miles per kWh during 3 seasons and city driving. We average about 4 miles per kWh overall. Including plenty of highway and winter driving. I Included all taxes, fees, delivery charges in my calculations and it costs us just under 10 cents per kWh to travel 4-5 miles.

That works out to be

- $1=40mi,
- $10=400mi,
- $100=4,000 miles,
- $1000 in electricity = about 40,000-50,000 miles.

The cost of electricity varies in the U.S. but, the overall average is 12.7 cents/ which is only a penny more than it was 10 years ago!

REDUCED WINTER RANGE

The range and efficiency are greatly reduced in the dead of winter when it's freezing balls outside. I roughly estimate to get about 2/3 range and efficiency from December through March but, it's still tons cheaper per mile than ICE cars. We'll guestimate $100 in electricity to get 2,500-3,000 miles *worst-case scenario.*

We were using the factory 110amp extension cord charger for the first 6 months we owned the car. This worked fine when we were only using a half tank of electrons and recharging all night. But, when the outside temps dropped and we used more battery range, the 110 cord didn't have enough time to recharge to max capacity. We purchased the factory recommended Bosch home charger for under $500. There are plenty of home charging options available including outdoor weatherproof chargers and build your own for under $100.

There were a couple of times that we drove around more than expected, suffered range anxiety, and turned the heat off to make it home. Keep in mind that this is a 2017 model and the real-world range is about 125-140 miles during 3 seasons. VW rates it at 126 city/111 highway. Most 2020 electric vehicles have much more range than this. We had to learn the hard way that we only get about 80-100 miles with the heat blasting. Now we are aware and don't take it further than 80ish miles per day during winter. No big deal, 80 miles per day is plenty for wifey. Most of today EV's have twice this range.

We can also program the car to preheat using the house electricity so we depart with full range and a warm interior. That's a smart feature that makes a difference.

THE GREEN FACTOR

We have also switched our electric supplier to mostly *renewable electricity* production as it feels silly to burn coal to charge my electric car. I have no idea why my electricity is supplied by coal when I live so close to the clean power of Niagara Falls.

I haven't done the math to recalculate cost per mile as I haven't noticed much difference in our bill and I am changing to 100% renewable energy soon. Jean, our children, and I all feel really good about emitting much less pollution.

People tend to associate electric cars with a political stance or opinion about climate change. The simple fact is this, if you run your car in a garage with the door closed you will die of carbon monoxide poisoning. I've inhaled enough car exhaust to last my lifetime. It feels good to minimize future exposure for, my family, the world, and myself.

There is also concern about mining for lithium. I did not research deep into this as I have a feeling that it isn't much worse for the earth than creating an extra 20,000 ICE parts from plastic, aluminum, magnesium, and other materials.

The automakers are already investing billions into responsibly recycling the batteries. Especially in Europe. The world will catch on.

THE DRIVING EXPERIENCE

Our electric car *is ultra-smooth and quiet.* When I carpool, my buddy who can't sleep in cars is gently swayed into a nap in the solid, quiet, comfy eGolf.

Here is a quick summary of the different driving attributes.

POWER

It has instant torque for fun acceleration at any moment. I love to hammer down and launch out of a corner. It is addicting in an EV.

ICE cars have to be at a certain RPM for best power, then mix air and fuel, compress it, and explode it, to create power. The power gets transfered through the crankshaft, through the transmission, before it gets to the driveline, and I get to use it through the tires on the road. While the process doesn't take long, I definitely notice the delay in the process after driving an electric car with no delay.

The eGolf isn't especially fast but, the instant torque is definitely fun. There is also a hidden "turbo button" under the accelerator.

If I am cruising in eco plus mode, need full power to enter the highway, but I already have it to the floor. I can push harder to find a delightful boost of full power without having to use the button on the counsel to select normal mode.

All EV's have instant torque, some have ridiculous amounts. Search You Tube for Tesla drag races and you'll be blown away by videos of ludicrous Teslas beating racecars, sportscars, and supercars down the drag strip.

HANDLING

The battery is under the floor so the low weight creates great cornering and stability. As a racer this is my favorite feature about all electric cars and trucks. Less weight is best but if you must have the weight it has to be as low as possible and between the axles. This is probably the most fun fwd car I have ever driven. It corners well, steers well, the adjustable regen can be engaging.

Low weight has a huge effect on handling and stability. So huge in fact that I'm dying to drive an electric truck. I'm guessing the low weight of the battery makes the best handling trucks and Suv's ever created.

Our eGolf is good in the snow. We got mama a set of Blizzaks. We run snow tires on all our whips. In her words, "Driving in snow is lovely. If you put it in regen mode you don't slide anywhere, you just drive. I only wish everyone else would get out of my way".

Electric cars also have the advantage of ultra-quick reaction through electronic stability control. The electric drive train can correct a slide or adjust power output much quicker than a traditional ICE drivetrain. This is a major factor in what makes electric vehicles so safe to drive.

ELECTRIC ENERGY REGENERATION

The Selectable regen is also fun to master. It seems to me that the most efficient technique is to coast forever then, at just the right moment slap the shifter down to max regen while approaching the red light or off-ramp. If you were too early, give it the tiniest amount of throttle to slow the deceleration. If you were too late use the brake. If you were dead on, soak in electrons and use the brake at the very end of the stop.

VW doesn't allow regen to bring the car to a complete stop, I'm guessing this is so we have to use the brake pads to clean surface rust off the barely used rotors.

This thing coasts like nothing I"ve ever felt! It was seriously the major "wow factor" for me. I was expecting the instant torque, silent sound, adjusting to regen driving, but no one mentioned the crazy coasting ability.

Think about it, no engine compression, no valve springs, no oil sloshing around spinning crank journals. No cam, and rod bearing friction. NO TRANSMISSION, fluid, clutches, gears, drums, a heavy torque converter, oil pump etc.

The only thing to slow it down is factory equipped low resistance tires, axle joints, and wind.

It seems that Porsche agrees coasting is most efficient over one pedal regen driving since the new Taycan reportingly coasts and uses regen only in its brake pedal.

UTILITY

Our eGolf is the exaxct same car as a regular ICE VW Golf. It is a smart utilitarian Hatchback design that the Germans have been perfecting for 40 years.

We take it everywhere! By we, I mean my family of 5 and sometimes our 100 lb Rottweiler lays or sits comfortably in the hatch. The ergonomics are incredible. We got it just in time to help move our stuff to our new home. With all the seats folded down Jean called it her "e-truck".

The max cargo factor was her major "wow factor". She is definitely not as concerned with the smart, efficient, reliable design as I am. She loves driving it, she loves saving money in fuel, and not polluting. However, she had no idea such a small car could fit so much stuff.

STYLING

In our shared opinion, the handsome design is the most attractive hatchback out there. The LED taillights look super sharp, the headlights are aggressively angled, the body lines are clean, and the proportions are good. The fit and finish are good both in and out. In typical German fashion the interior isn't flashy, cheesy, or busy. It's just done well with nice materials and all the buttons/controls where they should be.

There are other manufactures besides Tesla that are finally following suit. They finally realized that EV's don't have to look like a space ship took a shit and it landed on 4 wheels. I'm looking at you, Nissan Leaf and Mitsubishi i-MiEV. The i3 and Bolt are a little better but, still really nerdy looking.

To build an EV that is subtle and attractive is commonsense. I'm glad everyone else is now catching on.

QUALITY/VALUE

It doesn't feel like a cheap hatchback *it feels like a refined European car*. I already mentioned the excellent egonomics but, I have to note the excellent driving position. Something about it just feels right. I can drive this thing all day and not get worn out. It soaks up bumps nicely, the mateirlas are nice, the features are generous.

I feel that this car is an excellent value and an all-around winner. It's really good at everything it does and it does a lot. I tell my love, Jean it reminds me of her. Attractive, hard-working, surprisingly fun, super smart, modern, dependable, loving, and capable, all in a nice little size and shape.

Like dogs and cats, cars also take after the personality of their owner. We even start to look alike after a while. Pay attention and you'll see it sometimes.

Considering we ended up at $17,500 after trade-in and incentives for a brand new car with very low operation costs. I'd say this is an excellent value. Some people pay that for a new Ford Ecosport, Yuck!

I'd also like to note that our EV makes us smile every day. It's fun and easy to drive, smart, handsome, and efficient. I'd love everyone to figure out what makes them happy about a car and get that daily source of happiness. The eGolf definitely checks that important box for us.

ELECTRIC VS INTERNAL COMBUSTION

Since the electric car is ridiculously cheap to run my ICE cars stay in the driveway. They pollute the air my children breathe and cost us money to run.

I only drive my fuel burners when I have to or for fun at the race track. We take the Bimmer on long road trips, to haul the racing kart, and I drive it to work.

We take the ol' Burb when we are rolling 6 to 9 people deep, hauling the big trailer, or to get ice cream for the fun of it.

Other than that we take the EV everywhere. School run, church, grandma's, groceries, to get this or that. The Bimmer isn't bad at 25 mpg but, the *electric car just makes so much more sense, in stop and go driving.* Sometimes, if I have a lot of errands to run on my day off, I'll drop Jean to work and use her eGolf to run all over town, then pick her up later. It's worth it to not burn fuel all day sitting in suburban traffic. It just doesnt feel good to sit at redlights burning fuel while sitting still.

Jean had a vacation and I took the eGolf to work 5 days in a row. Afterword, I got in the Bimmer and it felt like an antique. I know it is now 19 years old but, it still has tighter steering, better brakes, and a smoother engine than most entry-level new cars. It's a European sports sedan that, for the first time, felt like 100-year-old technology.

It wastes energy in friction, noise, and heat. It feels sensless to run such a complex, inefficient, internal combustion machine after driving electric.

Jean and I used to switch cars sometimes when she had the Pathfinder. She used to like driving the Bimmer. Now, after driving electric for two years she complains that the Bimmer is a freaking work out and way too much effort to drive.

As much as I enjoy the driving dynamics of the Bimmer, I get her point. I personally don't mind putting in more effort as the driver but, I'm a driving enthusiast. Most people don't care to manually shift gears and feel the road through heavy steering.

The thing that bothers me is that the internal combustion machine itself has to use so much more effort to do the equivalent work of an electric machine.

Our electric car makes my Chevy V8 powered toys feel even more recreational. There are millions of V8's on American roads every day. This used to be the best way to get from point a to b but, that time is ending as we speak.

MOVING FORWARD

We will never buy another ICE car. The future is here. There are so many fully electric vehicles at so many different price ranges.

Jean is looking forward to a fully electric Range Rover. When the time comes we'll lovingly keep the eGolf in the family for our teen daughter to drive. It has a 10 year, 80,000-mile warranty on the battery against 20% loss. If we are getting 20% less range at 80K, I'll have it warrantied. If not our kids will drive it for all its worth. Either way we will get our money's worth out of it.

Car purchases are pretty much the worst investment one can make. It's the only thing that we spend thousands on that immediately plummets in value and costs us money to own.

Energy.gov reported in April of 2020 that The average household spent about $60,000 in 2018, with $9,800 of that on transportation (16%). Transportation expenditures were second only to housing (33%). The largest portion of transportation expenditures was for vehicle purchases. Gasoline and motor oil, public transportation, and other vehicle expenditures such as maintenance, repairs, insurance.

Our electric car will provide many years of note free driving and will be virtually free transportation for my daughter as her first car. Don't worry, I'll make her pay car insurance, phone bill, and ask her to run errands.

Today, transportation is the second-highest expense for an American household, after housing itself, and requires almost 30% of all the energy we use as a country.

It really does not make sense to buy a car today that burns fuel.

Electric cars are better in every way except for long road trips, hauling huge trailers, or just to have fun with vintage machines. So basically EV's make more sense for 90% of day to day travel.

Most people who take road trips and haul recreational campers/ trailers are family people with multiple vehicles. This presents a perfect opportunity to ease into EV ownership and still have all bases covered like our family does.

One vehicle that handles camping, road trips, hauling. Maybe a truck, minivan, or large SUV depending on your needs, hobbies, and preference. The other vehicle is electric and handles most of the families day to day needs. The EV goes on the long daily commute and runs all the daily errands. The Truck gets used as little as possible only when needed. At least until we can all get electric trucks. Yes I'm suggesting that parents or couples share vehicles to minimize transportation costs. Use the extra money for a vacation.

THE WORLD CHANGED. ELECTRIC CARS AND TRUCKS ARE SUPERIOR

The fastest cars in the world are now electric. I'm talking about roasting a Lamborgini in a huge Tesla SUV while hauling 4 buddies and their golf equipment. It is crazy what is happening right now in the transportation industry!

The biggest downfalls of conventional ICE Trucks and SUV's are eliminated by an electric powertrain. The lowered weight of batteries and super-efficient electric motor means that trucks and SUV's will no longer suffer from inefficiency and poor handling.

I'm super excited about fully electric trucks!

The Rivian is arguably the most capable thing ever built on 4 wheels. I'm talking about 0-60 in 3 seconds, hauling 11,000 lbs, crossing creeks, and climbing mountains. With 4 independent motors at each wheel it can perform tank turns, and provides excellent traction. It also has an independent air ride suspension to raise it over rocks or lower it down for excellent weight distribution which creates car-like handling and stability. All of this with 3 huge cargo areas adding up to more storage space than anything else on the road with roomy seating for 5! Did I mention that this is a vehicle that can do all that very well and go 0-60 faster than an Aston Martin DBS? My mind is blown!

Ford invested big money into Rivian to use the chassis for 2021 F-150s. America's best selling model for the last 36 years is going to be available as an EV with a huge front trunk accessible by folding down the front grill. Ford says it should be available late this year and has plans for 16 new EV's by 2022.

GM is bringing back the Hummer nameplate on an electric Pickup that will have 1000HP WTF!?! We are going to need special training to drive these monsters! The truck will be available in late 2021.

GM is now able to make electric vehicles equal cost or cheaper than ICE counterparts. It makes sense with approx 20,000+ less moving parts. GM has a new EV platform starting with Cadillacs and they are building the largest collective EV charging network in America. The new Caddie EV SUV looks amazing!

LOTS OF OPTIONS

Hyundai and Kia both have affordable EV Suv's and hatchbacks for sale right now. Along with 15 other models from different manufacturers ranging from $30,000-$100,000+

Don't worry, if you don't want to spend that much on a whip there are plenty of high value used models available.

- Used Nissan Leafs, go as cheap as $5k,
- Ford Focus EV $7k,
- BMW i3 $15k,
- Chevy Bolt $20k,
- Tesla Model S $30k

I'd argue that used is the best way to buy a Tesla since the federal rebate has been phased out on new Tesla purchases. https://www.fueleconomy.gov/feg/taxevb.shtml

VW is bringing the Microbus back on an all-new electric platform. They also have a fully electric Crossover and SUV coming out. The ID.4 SUV is expected late this year. VW group is planning to launch 27 new EV models by 2022!

VW, Volvo, Ford, GM, BMW, Jaguar, and most of China have solidified the EV as the future.

The Japanese automakers are finally deciding to catch up. Nissan has been around since the beginning with the Leaf but, Toyota didn't have an EV for sale in 2019.

All that is changing now as Honda and Mazda plan to fully electrify their fleets within 10 years and Toyota plans to get half its sales from EV's by 2025.

Tesla, who started this whole thing still makes the fastest accelerating, longest range, fastest charging, and smartest EV's out there. The current model S can get up to 370 miles per charge. They are approaching road trip use.

Tesla executed a brilliant plan by building the world's fastest luxury sedan (Model S) and SUV (Model X). They get our attention with this mind-blowing technology then roll out the affordable Model 3 sedan and Model Y Suv. They also have a fully electric Semi Truck and a Pickup truck for sale soon. All of these vehicles lead their respective classes and seem to stay 7 years ahead of all other   manufactures.

There are so many cool, fun, happy, SENSABLE, attractive EV,s coming out now. We are witnessing unprecedented changes in automotive history. Nothing this big has happened since transitioning from the Horse and carriage.

MY DREAM!

I'd love to convert old boxy Chevys, old Broncos, old Honda Civic hatchbacks, 90s Cherokee's, old VW Beetles, Land cruisers, and Range Rovers into fully electric.

The possibilities of tasteful transformations are endless!

Super efficient and low maintenance with no fuel or engine repair ever. Cool cars with the weakest link (inefficient, unreliable engines) addressed. A Modern electric motor in a lightweight classic. Perfect!

I feel like these cars would sell as fast as we could build them. It just makes too much sense and it is too fun not to catch on.

HARNESS COSMIC LIFE-GIVING ENERGY

We could have solar on EVERY roof to charge all of our electric cars and trucks. From the city houses to the country pole barns and the whole suburbs in between.

Tesla's solar roof tiles are beautiful and super smart when paired with a Tesla wall battery. If not tiles, regular panels work too. I envision solar on every rooftop exposed to SUN. Wind, hydro, geothermal energy works too.

We could have industrial solar. I'm not suggesting the solar farms that are popping up in nature. I'm talking about Solar roof over EVERY huge parking lot, ALL shopping plazas, every Home Depot, Wall Mart, and government building.

We could power the worlds the way God designed it and Nikola Tesla visioned it. If we invest now, the next generation will be that much freer to pursue other life. Solar is now the cheapest form of new energy in 60 countries.

We could even put solar on enclosed trailers connected to a Tesla battery mounted on the trailer which could add some range to the electric trucks hauling them. This couls also work on semi-trailers to boost the electric semis range.

WHAT ABOUT BUSINESS VEHICLES?

How much money will small businesses save with electric work vans, electric pickups, and electric box trucks? Landscapers, contractors, carpet cleaners, delivery service, etc. All small businesses that drive from job to job every day could save a lot in fuel costs.

What about the huge corporations of UPS, FedEx, Amazon, etc. They could reduce TONS of $Co2$ emissions by converting to full electric.

AND PUBLIC SERVICE

All government vehicles, mail trucks, city busses, can be electric and all government buildings and parking lots can be covered in solar.

Electric vehicles and solar panels would save tons in the prison system alone! Imagine the possibilities in all tax-funded entities and huge corporations.

WHY DOESN'T THIS EXIST NOW?

Let's do it!

My Dream continued

I'd love to Someday have a Rivian S and my fun fuel-burning Porsche 911.

An old classic Beetle transformed to fully electric for my oldest daughter. She thinks they are absolutely adorable.

My middle daughter will have the eGolf to complete the 3 generations.

The 911, Beetle, and Golf are all directly related. I'm no V-Dub fanboy but, it's only proper that we have the complete bloodline if those are our individual favorite cars.

We have to have our daily source of driving happiness.

Wifey can upgrade to the Range Rover she always dreamed of when they finally become fully electric. If not she really likes the Volvo XC90 too. Either way we are waiting for full electric versions.

My youngest RacinMason loves GM G-bodies. 1978-87 Monte Carlos, Malibus, Buick Regals, Olds Cutlass, etc. I know every nut and bolt on those cars as I spent about 10 years stripping, modifying, cutting, building, destroying, rebuilding them in the Street Stock racing division during my teens and 20s. I'll give him the choice if he wants to hear a V-8 Rumble for the fun of it or if he wants to convert it to fully electric with the Chevy electric crate motor coming out. Right now he is 9 and wants to make noise. I'm 37 and I get it.

POWER IT ALL CLEAN.

I'd love to have beautiful Tesla solar roof tiles covering my home, garage, shed, deck, and a solar carport in the driveway. All connected to a couple of Tesla power-wall batteries that manage my share of electrons as efficiently as possible.

Smart efficiency is so satisfying. I dream of charging all of our joyous transportation choices, and meeting my home power needs with nature.

I know the grid is a huge unknown but, if we can produce the energy we use, this naturally lightens the load on the grid.

Zero-energy homes are actually affordable. Many consumers, builders, and policymakers are reluctant to consider zero-energy homes because of the perception that costs are expensive.

In Detroit, a 2,200-square-foot net-zero energy house would cost $19,753 more than the same house with no solar and typical efficiency. The energy-bill savings would be $2,508 in the first year, and the solar and efficiency costs would pay for themselves in about nine years with inflation and other changes taken into account The Tesla power wall batteries will also be there to keep us going in case of grid failure. The initial investment will pay for itself. It's only a matter of time.

Heck I'd love to install solar on my enclosed race trailer too. It sits outside 24/7 and I could use a battery instead of loading, fueling, and listening to a generator at the race track or campground.

I'm itching to someday plug my "fully Charged" trailer into my

Rivian S hauler for added range, to farther race tracks. Not that I would need it but, smart, sensible, tasteful, efficient design and use of natural energy are so satisfying. It just feels good.

Some may say I'm dreaming. Perhaps I am. But, this is the dream I am working toward. I expect to experience this within my lifetime. The technology is here. The time is now. Let's embrace it.

Thank You for reading!

I have been passionate about cars, design, and function for 30 years. I hope you feel that you gained some understanding and awareness in the automotive world.

I also write about gaining understanding and awareness to use our personal power in different areas of life. Including but not limited to

Parenting, Relationships, family finances, and overall health in mind, body, and spirit.

Check out my blog posts at https://www.BobbyWes.com

https://www.instagram.com/BobbyWes33/

https://twitter.com/bobbywes33

https://www.facebook.com/Personal-
Power-106316151026246

www.ingramcontent.com/pod-product-compliance
Lightning Source LLC
Chambersburg PA
CBHW031500210526
45463CB00003B/1013